Fluid Power
Educational
Series

Concepts of Load Sensing Hydraulic Systems (In the English Units)

Joji Parambath

Concepts of Load Sensing Hydraulic Systems
(In the English Units)

Copyright © 2023 Joji Parambath

All rights reserved

ISBN: 9798854249164

https://jojibooks.com

Disclaimer of Liability

The contents of this book have been checked for accuracy. We cannot guarantee full agreement since deviations cannot be precluded entirely. Only qualified personnel should be allowed to install and work hydraulic equipment. Qualified persons are those who are authorized to commission, ground, and tag circuits, equipment, and systems following established safety practices and standards.

Table of Contents

PREFACE

When designing hydraulic systems, you aim to get as much work done as possible while using less energy. You've got many design choices - from more traditional circuits to specialized arrangements like load sensing. High-end load sensing hydraulic systems offer better control and energy efficiency. Load sensing systems have become increasingly popular and widely accepted, introducing greater sophistication and simplicity in hydraulic controls.

The book is really helpful when it comes to understanding load sensing hydraulic systems. It explains these systems' important components, like load sensing variable-displacement pumps and pump displacement control elements. When the system demands flow, a load sensing variable-displacement pump with a swash plate design delivers only the flow required by the associated system at a pressure required to move the connected load. The displacement control elements, including pressure controllers and load sensing compensators, are used to control the swash plate angle efficiently. For easy understanding, the book provides simple examples that clearly illustrate different modes of operation for load sensing systems. The English system of units is employed throughout the book.

The author gives many other topics in books under the fluid power educational series. A list of all the textbooks is given at the end of the book. Also, please see the details at https://jojibooks.com.

Enjoy reading the book.
Your feedback is most welcome.

JOJI Parambath

Chapter 1 | Introduction to Load Sensing Systems

In a hydraulic system, a pump is designed to deliver the necessary power to meet the load demands in the system.

A conventional hydraulic system with a fixed-displacement pump provides a constant fluid flow. The pressure in the system is developed due to the resistance offered to the flow. The load mainly imposes this resistance. A pressure relief valve (PRV) limits the maximum system pressure. When the system develops the set pressure, the pump flow is bypassed through the PRV, consuming excess power and generating tremendous heat.

One way to minimize losses is utilizing a variable-displacement pump that reduces flow to the minimum level once the set pressure is attained. Even when a variable-displacement pump is employed, the system produces considerable heat.

However, a load sensing system provides only the flow and pressure as required by the connected load. Implementing a load sensing system minimizes the need for multiple components and extensive plumbing, resulting in a simplified design, greater flexibility, and improved reliability.

The following section explains the characteristics of the conventional hydraulic systems.

The Behavior of Conventional Hydraulic Systems

In the conventional hydraulic system with a constant-displacement pump, the pump always discharges the flow at a constant rate. If the system requires no flow, the excess flow will have to be dumped into the associated tank, which is an inefficient waste of energy.

In the alternative variable-displacement design of the pump, the flow rate can be varied depending on the load requirements of the system.

Next, the circuit representations (Figure 1.1) and characteristics of several typical hydraulic systems are given below for a comparative study.

A system with a fixed-displacement pump

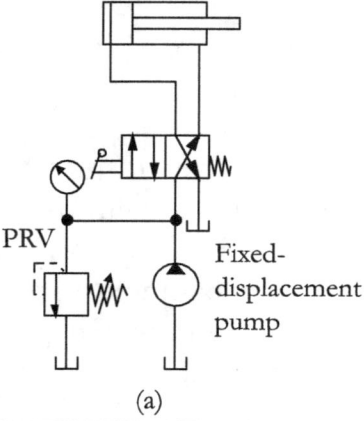

PRV

Fixed-displacement pump

(a)

A system with a variable-displacement pump and pressure controller

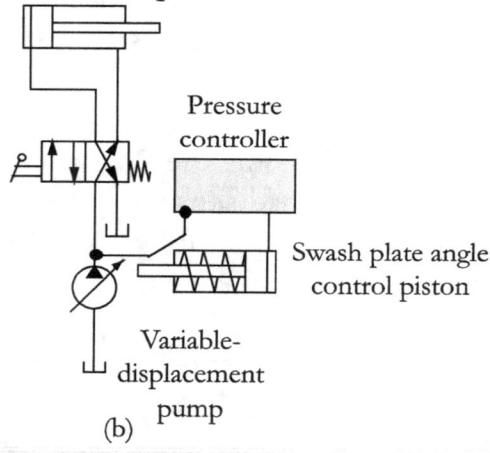

Pressure controller

Swash plate angle control piston

Variable-displacement pump

(b)

A system with a variable-displacement pump, load sense compensator and pressure controller

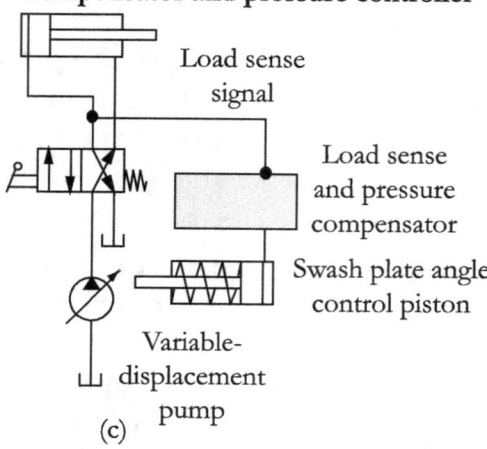

Load sense signal

Load sense and pressure compensator

Swash plate angle control piston

Variable-displacement pump

(c)

Figure 1.1

A Hydraulic System with a Fixed-displacement Pump

Consider a hydraulic system with a fixed-displacement pump for delivering a constant flow 'Q_s', a pressure relief valve (PRV) for limiting the maximum system pressure 'P_s', and a connected load demanding a flow 'Q_L' at the load pressure 'P_L'.

In this system, when the load continuously uses all the pump flow for the conversion to useful work, the load entirely uses the power delivered by the pump.

$P_s x Q_s = P_L x Q_L$, assuming there are no inherent inefficiencies due to leakage and pressure drops. Figure 1.2(a) gives the characteristics showing the power utilization.

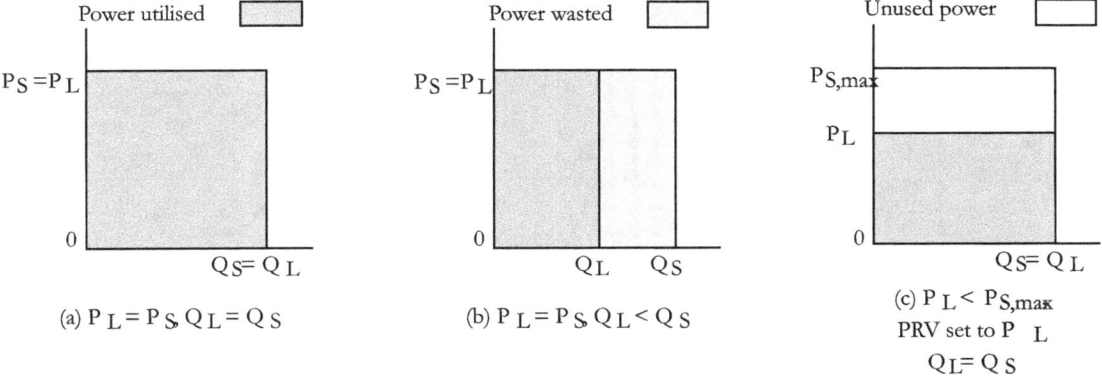

Figure 1.2 | P-Q characteristics of a hydraulic system with a fixed-displacement pump.

Next, when the load demands less flow than that delivered by the pump ($Q_L < Q_S$), the load utilizes only a partial amount of the power delivered by the pump. The remaining flow passes through the PRV and is converted into heat, as shown in Figure 1.2(b).

Furthermore, when the load-induced pressure is less than the maximum pressure setting of the system ($P_L < P_{S, max}$), the system pressure can be adjusted to a value just enough to meet the load demands. The pump cannot utilize its full capacity, as shown in Figure 1.2(c).

A Hydraulic System with a Variable-displacement Pump

When a hydraulic system is fitted with a variable-displacement pump, the load utilizes only a part of the available flow at less than the maximum system pressure.

Because this type of control regulates the pump flow at the pressure setting of the pressure controller, power is lost due to the potentially significant pressure drop in the system.

Figure 1.3(a), (b), and (c) gives the characteristics of power utilization.

Figure 1.3(a) shows the case when all the pump flow is used for a maximum pressure demand.

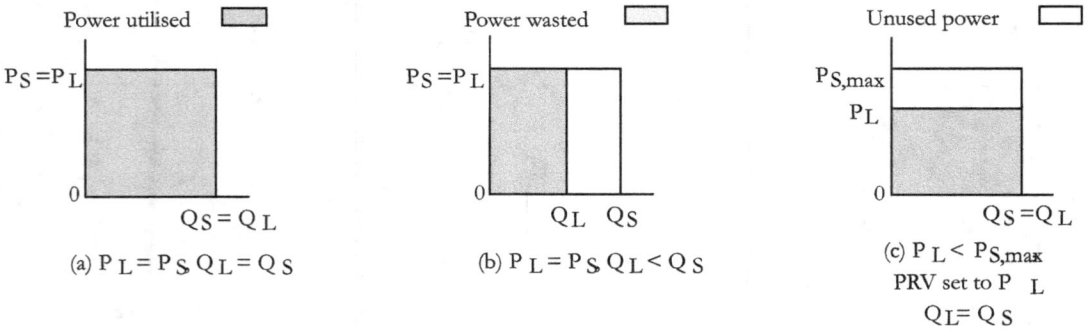

Figure 1.3 | P-Q characteristics of a hydraulic system with a variable-displacement pump.

Figure 1.3(b) displays the case when the system utilizes less flow during the approach of the maximum pressure demand. It may be noted that when the set pressure is reached, the flow tends to become a bare minimum.

Figure 1.3(c) shows the case when all the pump flow is used for less pressure demand by setting the controller to a pressure of P_L.

A Hydraulic System with a Load-sensing Feature

Figure 1.4(a) shows the characteristics of a conventional hydraulic system with a fixed-displacement pump, utilizing less flow and less pressure than pump-rated values.

Figure 1.4(b) shows the characteristics of a hydraulic system with a variable-displacement pump, utilizing less flow and pressure than pump-rated values.

Ideally, the hydraulic system should provide only the flow and pressure as the connected load requires.

Conventional hydraulic systems with constant and variable-displacement pumps cannot offer these ideal features most efficiently, and only a load sensing (LS) system can approach the capabilities of the ideal hydraulic system. Figure 1.4(c) shows the characteristic of the load sensing system.

The following examples illustrate the power generation, utilization, and efficiencies of different hydraulic systems in response to the applied loads. The latter sections explain the operation of the load sensing system.

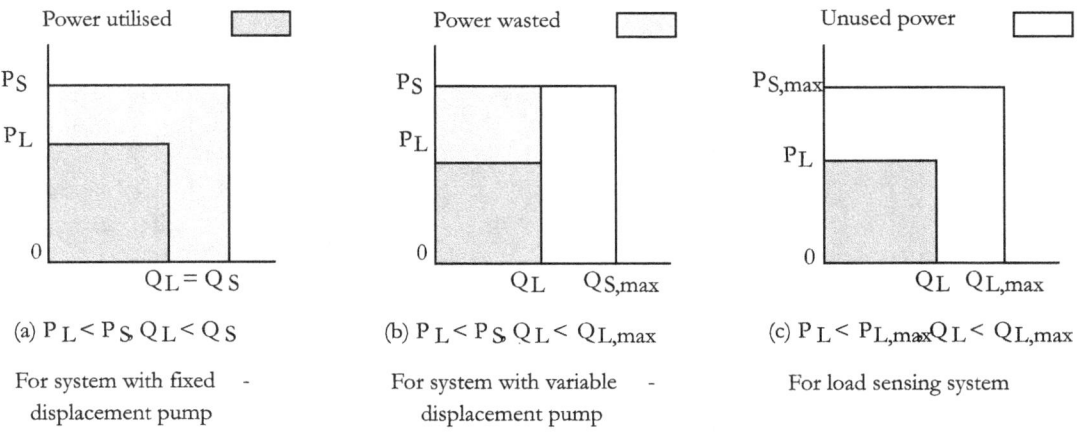

Figure 1.4 | P-Q characteristics of hydraulic systems utilizing less flow and less pressure.

Example 1.1 | A conventional hydraulic system with a fixed-displacement pump delivers 10.6 gpm and drives a load through a cylinder requiring 2.64 gpm at a pressure of 1450 psi, as shown in Figure 1.5.

The system pressure of 4350 psi is set by using a PRV.

A closed-center 4/3-DC valve is used as a final control element to control the cylinder.

Calculate (1) the power delivered by the pump in the neutral position of the valve, (2) the power utilized by the load, (3) the total power delivered by the pump during the extension stroke of the cylinder, and (4) the system efficiency.

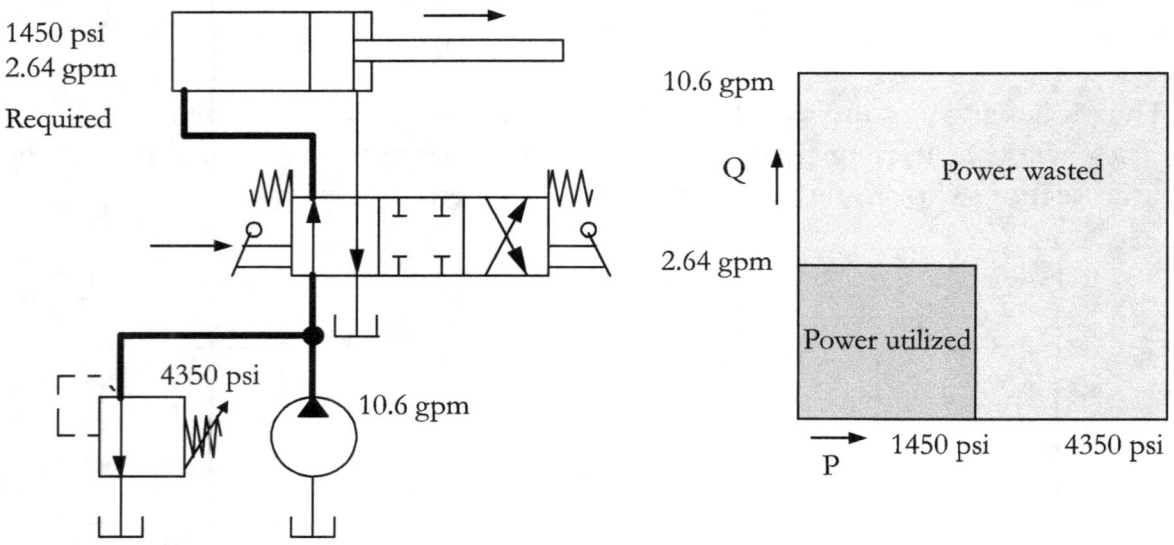

Figure 1.5 | A conventional hydraulic system with a fixed-displacement pump and its P-Q characteristics. (Example 1.1)

Solution

Pump delivery, Q	= 10.6 gpm
Required load flow, Q_L	= 2.64 gpm
System pressure, P	= 4350 psi
Load pressure, P_L	= 1450 psi

Power delivered by the pump when the valve remains in its neutral position

$$= PxQ/1714$$
$$= 4350x10.6/1714 \text{ hp}$$
$$= 26.9 \text{ hp}$$

Power drawn by the load when the valve is shifted to its left envelope

$$= 1450x2.64/1714 \text{ hp}$$
$$= 2.23 \text{ hp}$$

Total power delivered by the pump at the end of the extension stroke of the cylinder

$$= 4350x10.6/1714 \text{ hp}$$
$$= 26.9 \text{ hp}$$

System efficiency (end of stroke) = (2.23/26.9)x100%
$$= 8.3\%$$

Inference: The system efficiency tends to be too low if the load requirement of pressure and flow is lower than the system (pump) rating.

Example 1.2 | A hydraulic system with a fixed-displacement pump delivers 10.6 gpm and drives a load through a cylinder requiring 2.64 gpm at a pressure of 1450 psi, as shown in Figure 1.6.

The system pressure of 4350 psi is set by using a pressure relief valve.

The double-acting cylinder, controlled by a closed-center 4/3-DC valve, must be unloaded at the ends of its strokes with a standby pressure of 145 psi.

Calculate: (1) the power delivered by the pump in the neutral position of the valve, (2) the power utilized by the load, (3) the total power delivered by the pump during the extension stroke of the cylinder, and (4) the system efficiency.

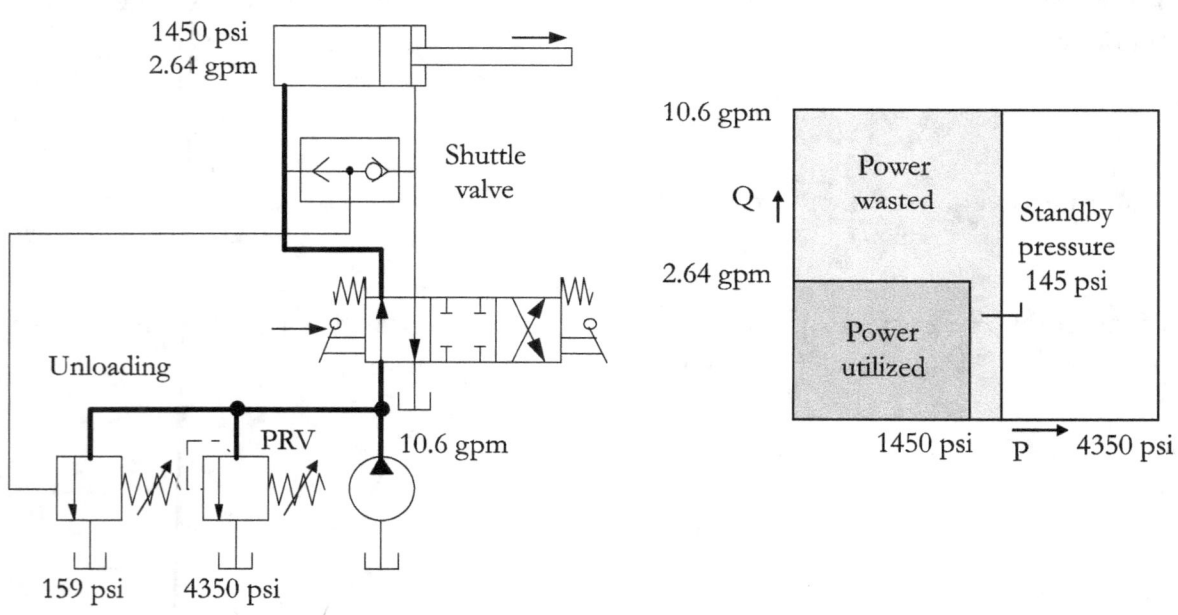

Figure 1.6 | A hydraulic system with a fixed-displacement pump unloading at low standby pressure and its P-Q characteristics. (Example 1.2)

Solution

Pump delivery	= 10.6 gpm
Required load flow	= 2.64 gpm
System pressure	= 4350 psi
Load pressure	= 1450 psi
Standby pressure	= 145 psi

Power delivered by the pump in its neutral position

$$= PxQ/1714 \text{ hp}$$
$$= 145x2.64/1714 \text{ hp}$$
$$= 0.223 \text{ hp}$$

Power utilized by the load = 1450x2.64/1714 hp
$$= 2.23 \text{ hp}$$

Total power delivered by the pump during the extension stroke of the cylinder
$$= 1595x10.6)/1714$$
$$= 9.86 \text{ hp}$$

System efficiency = 2.23/9.86x100%
$$= 22.6\%$$

Inference: The use of the unloading valve improves system efficiency.

Example 1.3 | A hydraulic system consisting of a variable-displacement pump delivering 10.6 gpm and pressure compensation setting of 4000 psi, driving a load through a cylinder requiring 2.64 gpm at 1450 psi, as shown in Figure 1.7.

The system pressure of 4350 psi is set by using a pressure relief valve.

Assume the leakage as 0.1321 gpm.

Calculate the (1) Power delivered by the pump in the neutral position of the valve, (2) Power utilized by the load, (3) Total power delivered by the pump during the extension stroke of the cylinder, and (4) System efficiency.

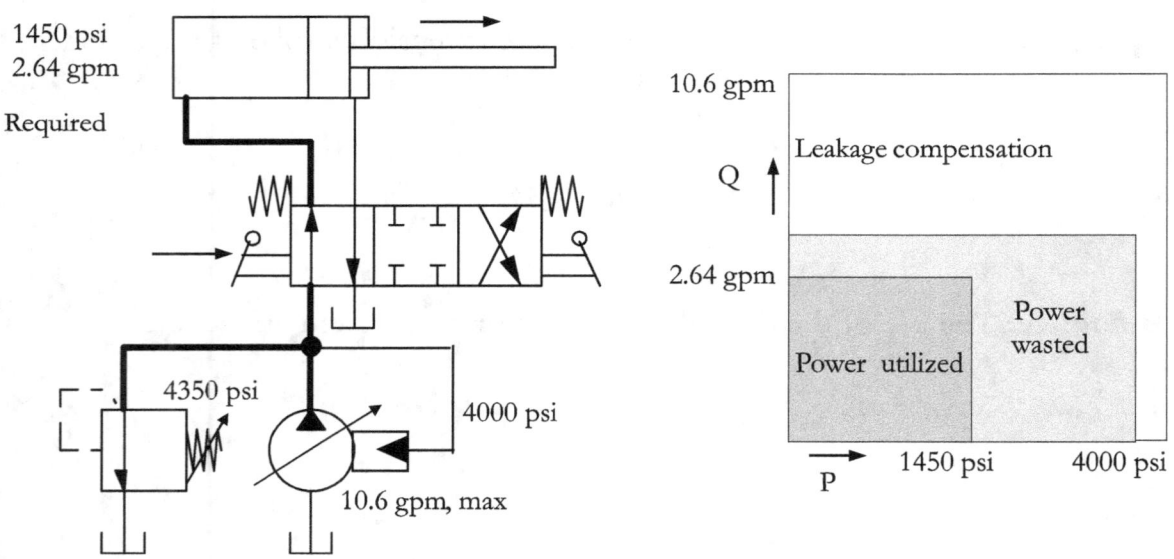

Figure 1.7 | A hydraulic system with a load-sensing variable-displacement pump and its P-Q characteristics. (Example 1.3)

Solution

Pump delivery	= 10.6 gpm
Required load flow	= 2.64 gpm
Leakage	= 0.1321 gpm
System pressure	= 4350 psi
Pressure compensation setting	= 4000 psi
Load pressure	= 1450 psi

Power delivered by the pump in its neutral position

$$= 4000 \times 0.1321 / 1714$$
$$= 0.31 \text{ hp}$$

Power utilized by the load

$$= 1450 \times 2.64 / 1714$$
$$= 2.23 \text{ hp}$$

Total power delivered by the pump during the extension stroke of the cylinder

$$= 4000 \times 2.64 / 1714$$
$$= 6.16 \text{ hp}$$

System efficiency $= (2.23/6.16) \times 100\%$
$= 36.2\%$

Inference: Using the variable displacement pump further improves the system efficiency.

Example 1.4 | A hydraulic system with a variable-displacement pump delivering 10.6 gpm and with a standby pressure setting of 290 psi, driving a load through a cylinder requiring 2.64 gpm at 1450 psi, as shown in Figure 1.8.

The system pressure of 4350 psi is set by using a pressure relief valve.

Assume the leakage as 0.1321 gpm.

Calculate the (1) Power delivered by the pump in the neutral position of the valve, (2) Power utilized by the load, (3) Total power delivered by the pump during the extension stroke of the cylinder, and (4) System efficiency.

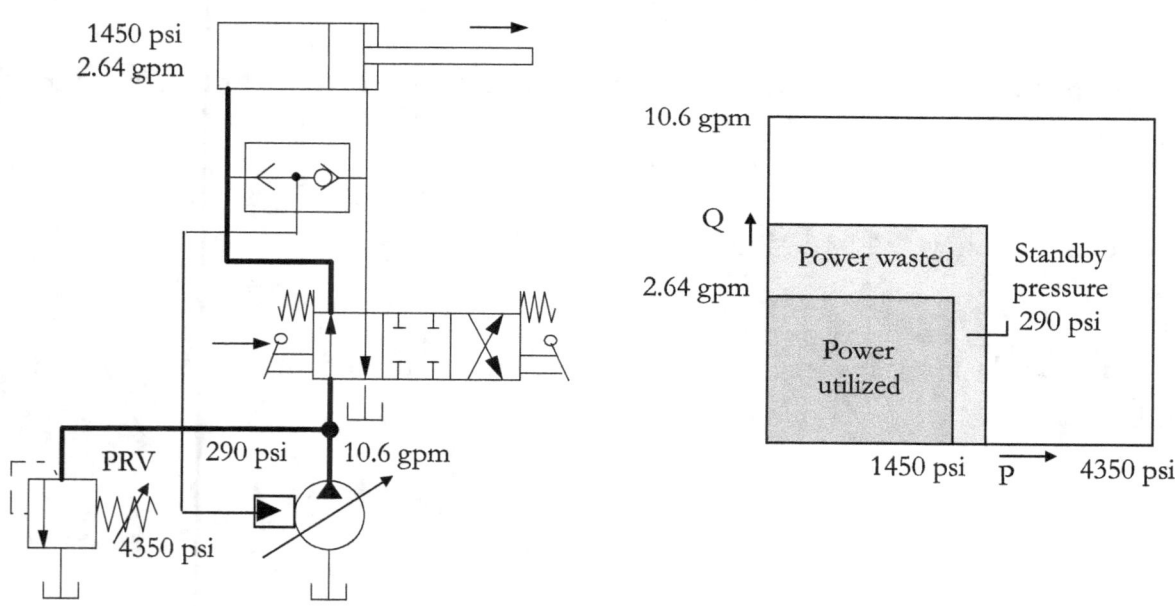

Figure 1.8 | A hydraulic system with a load sensing variable-displacement pump unloading at low standby pressure and its P-Q characteristics. (Example 1.4)

Solution

Pump delivery	= 10.6 gpm
Required load flow	= 2.64 gpm
Leakage	= 0.1321 gpm
System pressure	= 4350 psi
Pressure compensation setting = 290 psi	
Load pressure	= 1450 psi

Power delivered by the pump in its neutral position
$$= (4350 \times 0.1321)/1714$$
$$= 0.335 \text{ hp}$$

Power consumed by the load $= (1450 \times 2.64)/1714$
$$= 2.23 \text{ hp}$$

Total power delivered by the pump during the extension stroke of the cylinder
$$= (1740 \times 2.77)/1714$$
$$= 2.81 \text{ hp}$$

System efficiency $= 2.23/2.81 \times 100\%$
$$= 79\%$$

Inference: The use of the load sensing system further improves system efficiency.

Chapter 2 | Components of Load Sensing Systems

The load sensing system only produces just enough pressure and flow as demanded by the connected load in the system. It typically comprises a variable-displacement load sensing pump fitted with a special compensator and a load sensing directional control valve with proportional flow characteristics to supply the necessary flow to the associated system. The following sections explain the details of the components required for the load sensing system.

Load sensing Variable-displacement pump

Figure 2.1 shows the schematic and symbolic representations of the variable-displacement axial piston pump of the swash plate design. It consists of a group of cylinders with pistons arranged in a housing, a swashplate, a stationary valve plate, a driveshaft, a bias piston, a control piston, and a compensator valve assembly.

The cylinders in an axial piston pump are arranged in parallel and formed into a round block about an axis. A prime-mover drives the cylinder block in an axial piston pump. The cylinder block assembly rotates with the driveshaft. The cylinder pistons are fitted to the swash plate through ball joint shoes. If the swash plate is angled, the rotating pistons move back and forth in their respective cylinder bores.

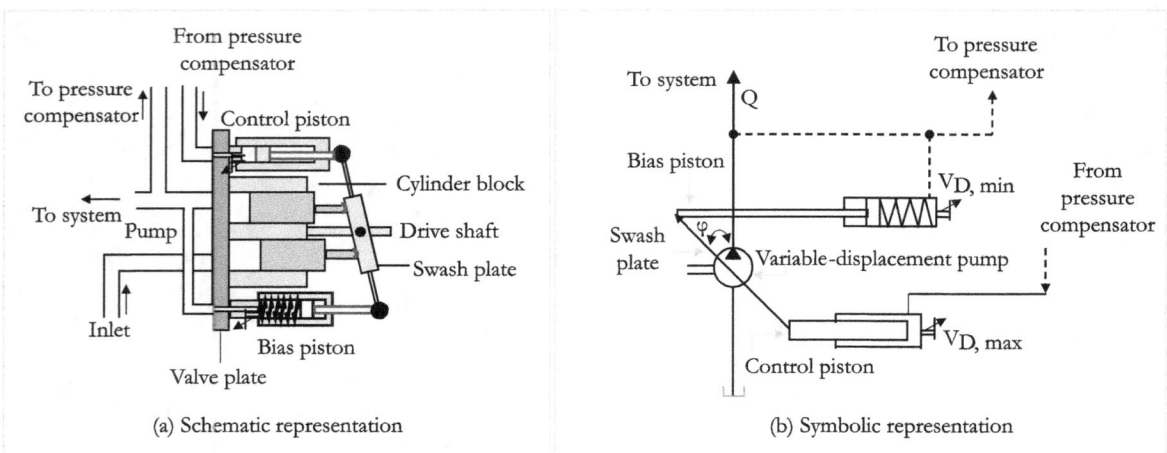

Figure 2.1 | Schematic and symbolic representations of a variable-displacement axial piston pump.

The pump flow can be controlled by varying the angle of its swash plate. This type of control can be achieved with the help of the control piston of the swash plate. The pump, along with its compensator, senses and responds to the associated hydraulic system's varying flow and pressure requirements in a manner as explained in subsequent sections.

Mechanical Minimum and Maximum Swivel Angle Limitation

The variable displacement pump has mechanical stop screws in the bias cylinder ($V_{D,min}$) and displacement control cylinder ($V_{D,max}$) to limit the minimum and maximum displacements.

Displacement (Maximum): The maximum volumetric displacement of the pump can be adjusted by using a mechanical stop screw called $V_{D, max}$ in the displacement control cylinder. Turning the screw inward will decrease the pump displacement while turning it outward will increase it in a typical case.

Displacement (Minimum): The minimum volumetric displacement of the pump can be adjusted by using a mechanical stop screw called $V_{D, min}$ in the bias cylinder. Turning the screw inward will increase the pump displacement, while turning it outward will decrease it in a typical case.

Classification of Axial Piston Pumps

Axial-piston pumps can be classified as in-line axial-piston pumps and bent-axis axial-piston pumps. In the in-line axial-piston pump, the center line of its cylinder block is arranged in line with the center line of its driveshaft. In the bent-axis axial-piston pump, the center line of its cylinder block is at an angle with the center line of its driveshaft. Load sensing hydraulic systems can utilize either type of pump.

Pump Displacement Controls

In a hydraulic system with a variable displacement axial piston pump, it is necessary to control the pump displacement to vary the flow rate through the system and limit the pressure in the system to the design value. Different methods can vary the displacement of an axial piston pump. Some of the important methods associated with pump displacement controls, like the pressure controller (limiter) and the pressure controller with load sensing, are described in the following sections.

Pressure Controller for a Variable-displacement Axial Piston Pump

We know that a high-pressure relief valve can control the pressure in a hydraulic system, but it also wastes much energy in the form of heat. To prevent this, a pressure controller (also known as a pressure compensator) with a servo system is often used in a system with a variable-displacement pump. This helps regulate the pressure without generating too much heat in the system. Figure 2.2 shows a cross-sectional view and a symbolic representation of a pressure controller.

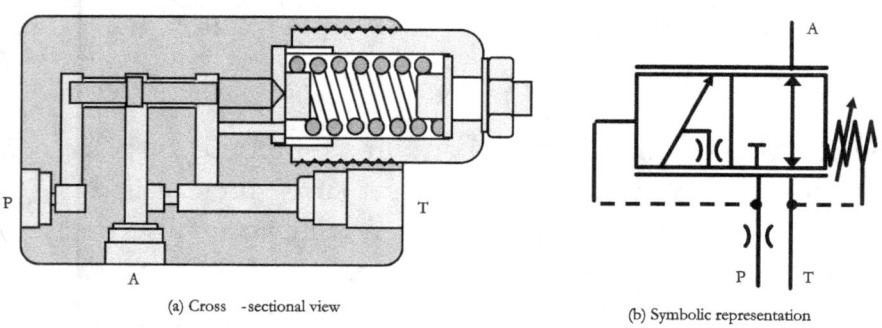

(a) Cross -sectional view

(b) Symbolic representation

Figure 2.2 | Pressure controller for variable-displacement axial piston pumps

The pressure controller with a servo system can effectively limit the maximum pressure at the pump outlet. The swash plate angle can be adjusted, and the pump displacement can be varied depending on the system pressure. This ensures that the pump delivers only the fluid the system needs.

When the pressure in the system goes up, the pressure controller sends fluid to the control piston to lower the swash plate angle and decrease the flow rate. Once the pressure reaches the maximum limit, the signal from the pressure controller moves the swash plate to its lowest angle position, decreases the flow rate to a minimum, and keeps the system pressure at the desired maximum level.

Variants of Pressure Controllers

There are pressure controllers that come with extra features. One example is a pressure controller with an extra port connecting an external pressure relief valve, allowing for remote pressure control. Additionally, there are pressure controllers designed for controlling the pressure of multiple axial piston pump units that operate in parallel.

Example 2.1 | Control of a Double-acting Hydraulic Cylinder Using a Servo-controlled Variable Displacement Pump

A double-acting hydraulic cylinder is used for high-pressure operation. The cylinder is powered by a variable-displacement axial piston pump and controlled by a 4/3-DC valve. To minimize energy loss in the form of heat, a servo-controlled pressure controller is required to set the pressure in the system. Develop a hydraulic circuit to implement the control task.

Solution

A Circuit with a Servo -controlled Variable -displacement Pump for the Control of a Hydraulic Cylinder

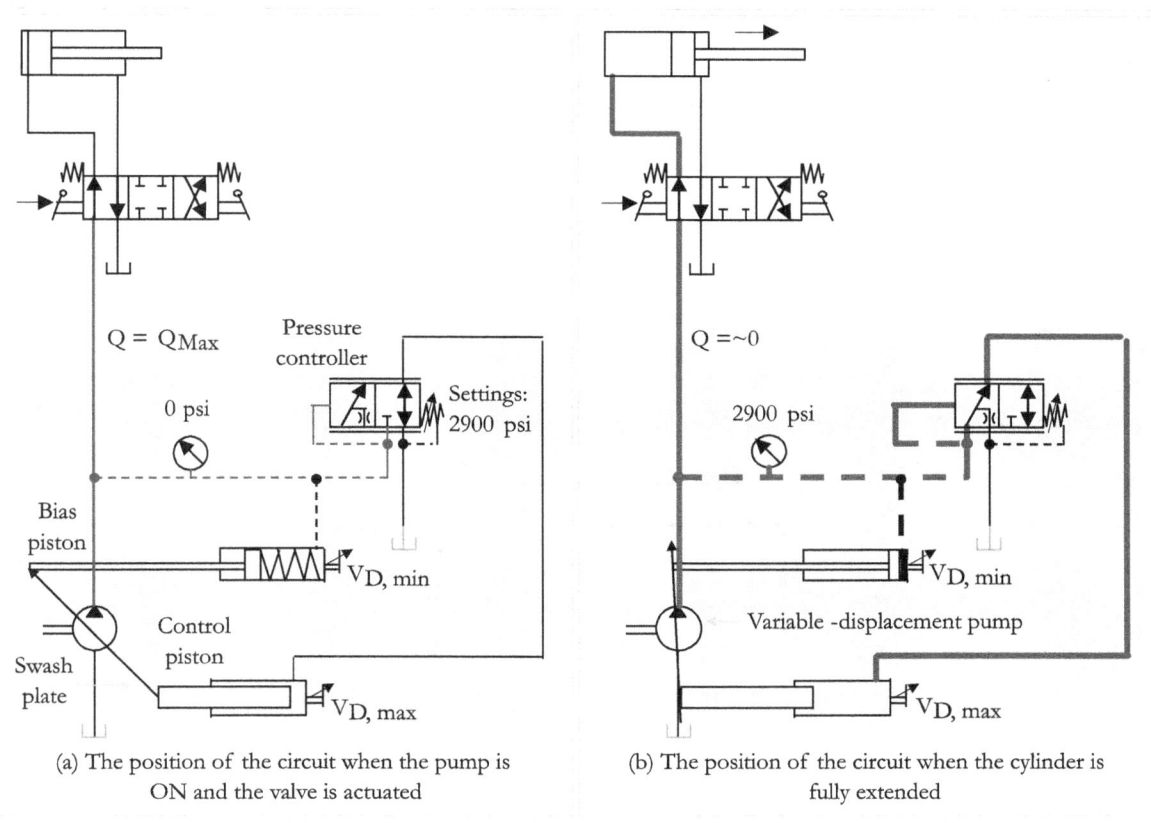

(a) The position of the circuit when the pump is ON and the valve is actuated

(b) The position of the circuit when the cylinder is fully extended

Figure 2.3 | Multiple positions of the circuit for the control of a double-acting cylinder powered by a variable displacement pump (Example 2.1)

The swash plate angle of the hydraulic pump with a variable displacement is controlled by the balance of forces produced by the bias and the control pistons. Further, the pressure controller can operate the control piston.

At first, the bias piston keeps the swash plate at its highest angle, allowing the pump to deliver the maximum amount of fluid. When the pump is turned on, it starts pumping fluid to the system, as well as to the bias piston and the pressure controller. The pressure controller then releases the fluid from the control piston to the tank. Figure 2.3(a) displays the circuit's position when the pump is turned on, and the 4/3-way valve is just actuated.

If there is any resistance to the flow, it can result in pressure building up at the outlet of the pump. This pressure acts on the left side of the pressure controller and works against the spring. As a result, the spool of the controller valve will move towards the right, opening the passage to the control piston and directing fluid in proportion to the pressure's magnitude. The control piston will then move and push the swash plate, which will decrease the flow output of the pump. Figure 2.3(b) shows the pump's circuit position when the cylinder reaches its end of stroke position, and the set pressure is generated in the system. In this position, the pump's output flow is just enough to supply the lubrication and leakage flows within the pump.

Static Characteristics of a Variable-displacement Pump

Figure 2.4 illustrates the static characteristics of a variable-displacement axial piston pump controlled by a pressure controller.

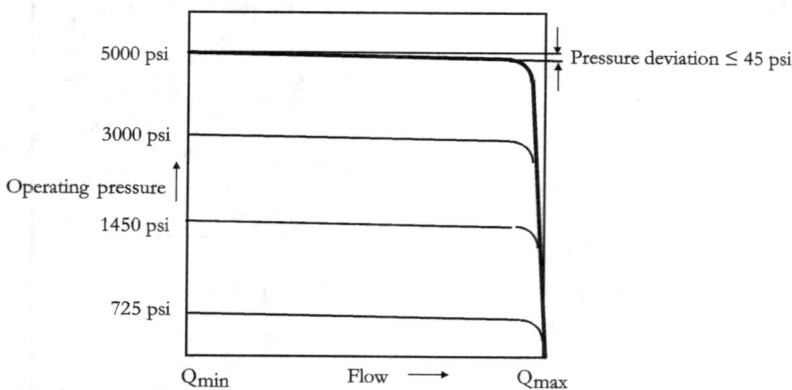

Figure 2.4 | Static characteristics of a variable-displacement pump

The diagram demonstrates a significant decrease in flow when the pressure increases in the final 45-psi band (from near maximum to minimum flow rate). However, when the pressure increases in the lower band, the pump flow only slightly declines.

Flow Control

A constant differential pressure (ΔP - Delta-P) created by an orifice installed between the pump and the cylinder can control the flow from the pump. The basic principle of Delta-P is described in the following section.

Basic Principles of Delta-P

Figure 2.5 shows the circuit of a hydraulic system used for lifting loads. The actuator for lifting the load is a double-acting cylinder. The power source used is a variable-displacement axial piston pump of the swash plate design.

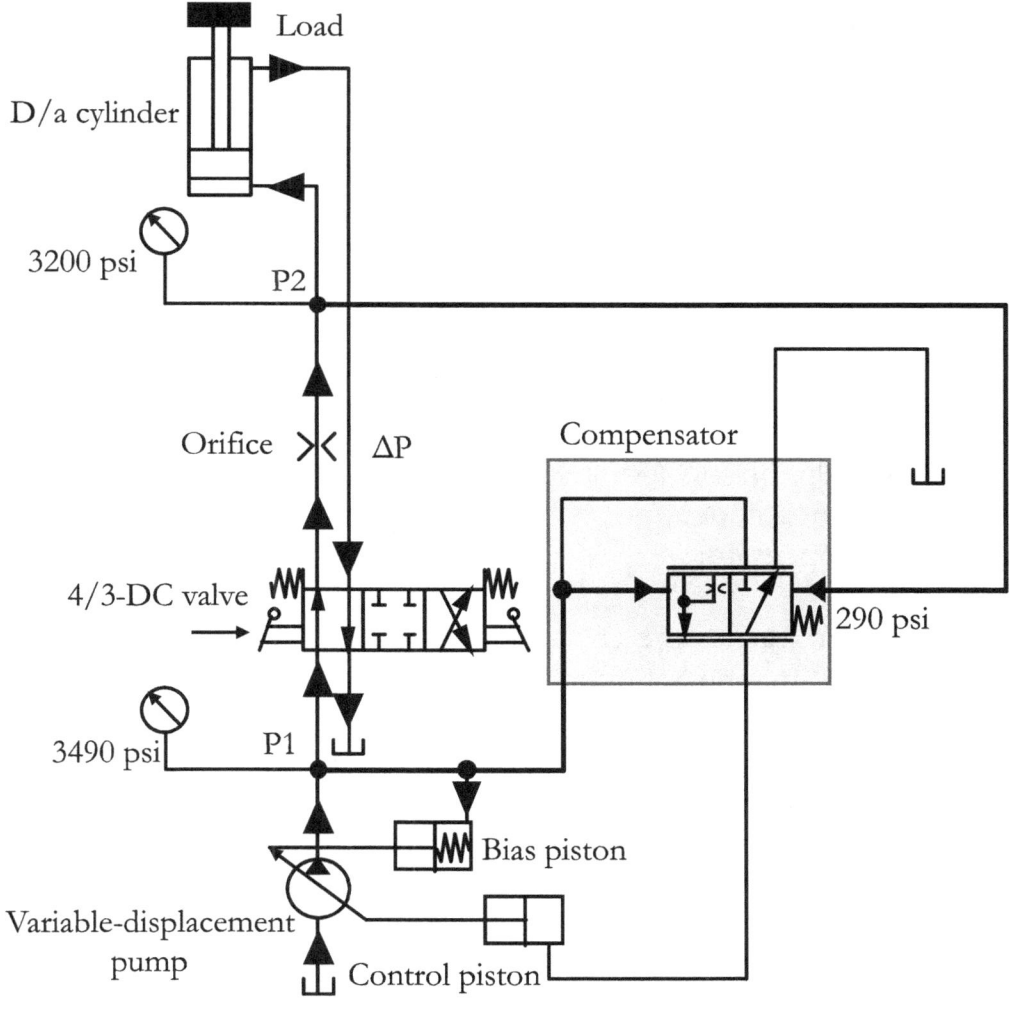

Figure 2.5 | A circuit of a hydraulic system used for lifting loads

The pump should be able to supply enough fluid to move the cylinder at the correct speed. A compensator with a load sensing feature can maintain a constant pressure differential across the orifice installed between the cylinder and the directional control valve. The load-induced pressure downstream of the orifice can be sensed, and the pump flow can be adjusted to maintain a constant pressure drop and flow across the orifice.

Forces on both sides act upon the servo spool of the compensator valve. On one side, the force due to pump outlet pressure P1 acts. On the opposite side, forces due to the 290-psi spring and the orifice outlet (load) pressure P2 act against the force due to pressure P1. That is, the piston is acted upon by the differential pressure Delta-P ($\Delta P = P1-P2$) and sends a corresponding signal to the control piston.

Next, the angle of the swash plate is controlled by a bias piston and a control piston. The bias piston tends to push the swash plate to create more angle and hence more flow with the help of the spring force and force due to the pump outlet pressure P1. In comparison, the control piston tends to push the control piston to create less angle and hence less flow with the help of the signal from the compensator, which is created by the force balance acting on the servo spool of the compensator. The pump will then supply the necessary amount of fluid with the help of the infinitely variable compensator valve to maintain Delta-P when the cylinder moves. This proportional valve can be called a pressure-flow compensator.

When the cylinder reaches the end of its stroke, the movement stops, and the pressure builds up. It is necessary to limit the maximum operating pressure to a safe value. A pressure controller called a high-pressure compensator or a pressure cut-off can be added to the pressure-flow compensator to limit the system pressure by reducing pump displacement to zero when the set pressure is reached. The pump supplies only the fluid required to maintain the maximum set pressure.

When all spools are in the center position, the load-signal port is vented to the tank, and the pump maintains 'standby' pressure equal to or slightly higher than the load sensing controller's delta P setting.

Pressure Transducers

A pressure transducer converts a pressure value into an analog electrical output signal, which the associated controller can use.

Swivel Angle Sensors

A swivel angle sensor measures the swivel angle of a pump's swash plate. The sensor with integrated electronics works on the Hall effect principle. It also has a sensitive probe tip.

Compensator with Load Sensing and Pressure Cut-off

Figure 2.6 shows the schematic and the symbolic representations of the compensator with load sensing and pressure cut-off. The compensator senses the pressure and flow conditions of the system. It consists of a pressure-flow compensator spool that works against a low-pressure spring (say 290 psi spring) and a high-pressure compensator spool that works against a high-pressure spring (say 4350 psi spring). The numerical values are used for easy understanding, and their real values depend on the actual hydraulic system concerned.

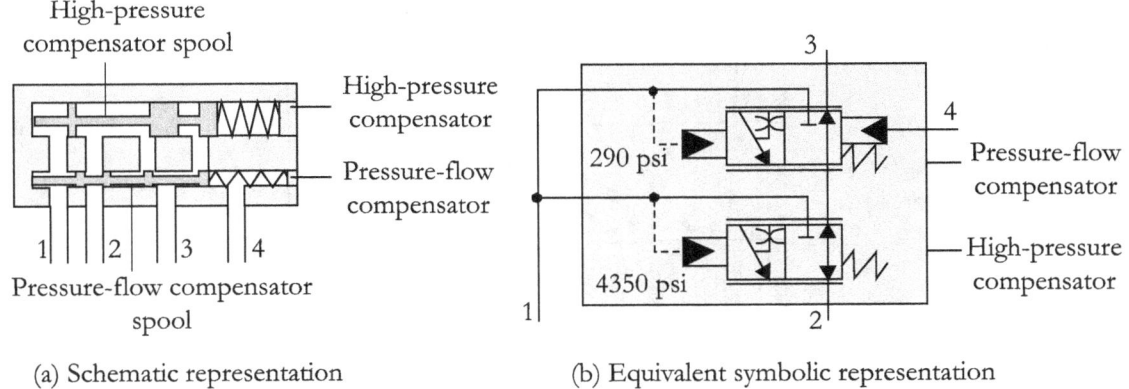

(a) Schematic representation   (b) Equivalent symbolic representation

Figure 2.6 | Schematic and symbolic representations of a compensator with load sensing and pressure cut-off features.

The pressure-flow compensator in a hydraulic system can sense the load-induced pressure downstream of an external orifice installed between the associated pump and actuator in the system through load sensing port 4, and the flow output of the associated pump can be adjusted to maintain a constant pressure drop across the orifice. The pressure-flow compensator compares the pressures upstream and downstream of the orifice and keeps the pressure drop over the orifice (ΔP) constant, thereby controlling the flow. An increase of ΔP causes the pump to de-stroke, and a decrease in ΔP results in a larger pump swivel angle till the flow control valve spool is in balance again.

The load sensing compensator is designed with proportional characteristics to control the variable-displacement pump to supply the amount of fluid required by the actuator.

The high-pressure compensator limits maximum operating pressure by de-stroking the pump, reducing pump displacement to zero when the preset maximum pressure is reached.

Characteristics, Pump with a Load Sensing Compensator
Figure 2.7 illustrates the static characteristics of a variable-displacement axial piston pump controlled by a load sensing compensator.

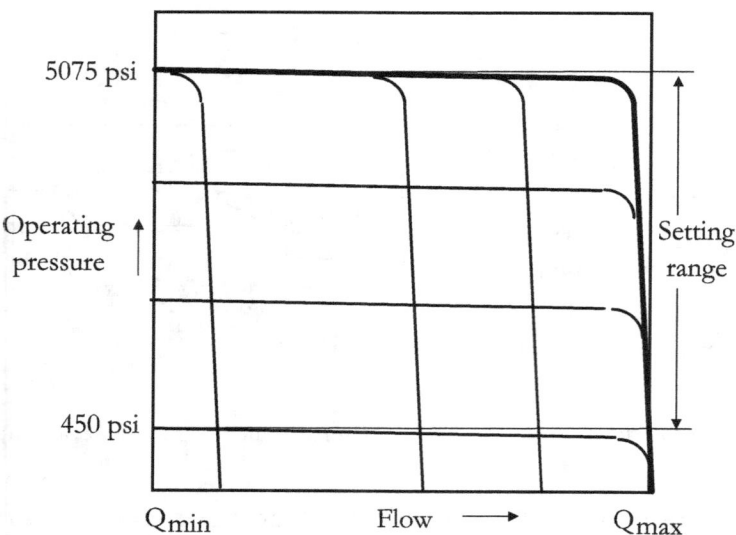

Figure 2.7 | Static characteristics of a variable-displacement pump

The diagram displays the flow behavior under various pressure settings. Additionally, it illustrates the flow behavior at the maximum pressure setting and different flow rate requirements.

Load-sensing Type Directional Control Valves

Using a load sensing type directional control valve in a hydraulic system increases machine performance, efficiency, controllability, and fuel saving compared to conventional directional control valves. The cross-section and symbol of a typical load sensing type closed-center 4/3-way directional control valve are given in Figure 2.8. The closed-center version is employed for systems with variable displacement pumps, whereas the open-center version is used for fixed displacement pumps.

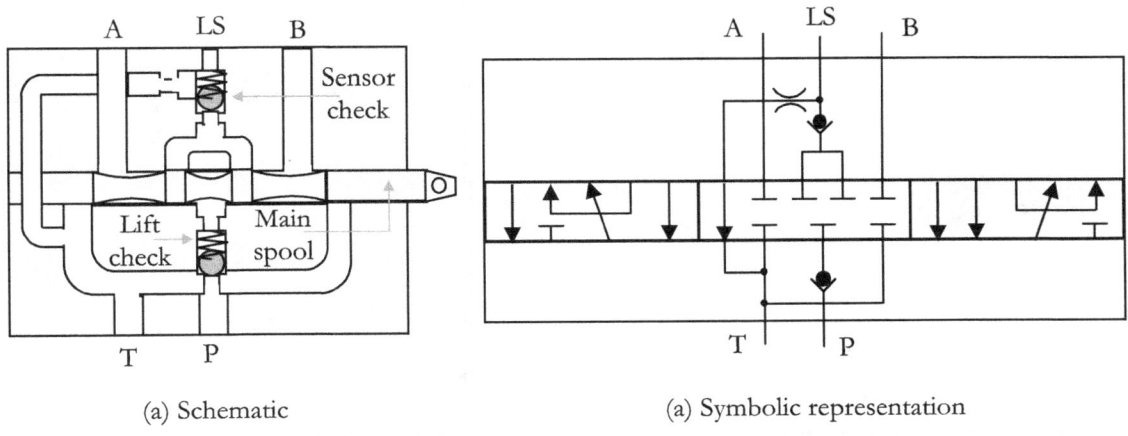

| (a) Schematic | (a) Symbolic representation |

Figure 2.8 | A typical load sensing type directional control valve

Usually, the valve has a load-holding check valve that permits the flow from the pump to the actuator while obstructing the flow from the actuator to the pump.

Additionally, the valve features a 3-way sensor check with an orifice. The check valve allows the flow from the pump to the compensator with negligible flow to the tank. It blocks flow from the compensator to the pump and redirects it through the orifice.

The valve can be operated manually, hydraulically, or electrically with on/off solenoids or proportional solenoids. A proportional system enables the main spool to be positioned infinitely, allowing for precise control of flow rate and actuator speed. The position of the main spool determines the rate and direction of flow through the valve.

When the spool in the valve is centered, pressure port P and working ports A and B are blocked. Actuation of the spool to one side can connect port P to port A and port B to port T. Moving it to the other side will connect port P to port B and port A to port T.

When the directional control valve is actuated, it produces a signal in accordance with the load pressure through port LS for the load sensing port of the related compensator. Additionally, the DC valve can create a path for fluid flow in the spring chamber of the compensator when it is centered.

Spool Stroke Limiters, Adjustable

The maximum opening of the main spool can be controlled by adjusting the stroke limits with adjustable limiters at either end or both ends of the valve. These limiters can restrict the maximum flow rate to the working ports A and B and be set mechanically.

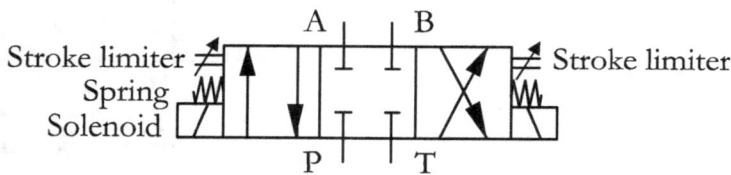

Figure 2.9 | Stroke limiters

Stacked DC Valves, Load Sensing Systems

Multiple load sensing directional control valves can be stacked to perform various functions, particularly in a mobile hydraulic system. The valve block can include numerous control features for load sensing.

Pressure Compensator, Load Sensing Systems

A pressure compensator can maintain a constant flow +to the actuator, even if the system pressure varies. The compensator adjusts for pressure fluctuations at the pump or working ports of the cylinder.

Anti cavitation Measures in Load Sensing Systems

In a load sensing hydraulic system, shock and anti-cavitation valves can safeguard components from pressure surges and cavitation. Remember, anti-cavitation valves can only prevent cavitation. A circuit with a relief valve and a check valve, as shown in Figure 2.10, can provide shock suppression and anti-cavitation features. The relief valve can release peak pressures by directing them to the tank, while the check valve can draw fluid from the tank to the related working port, thus avoiding cavitation.

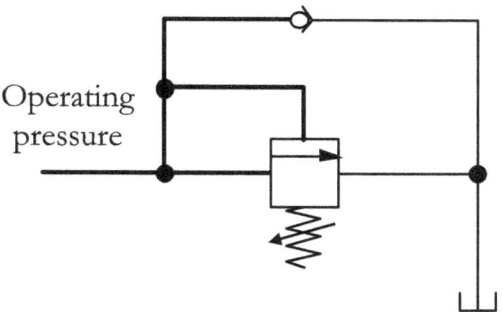

Figure 2.10 | Anti cavitation

Shuttle Valves, Load Sensing Systems

When using multiple load sensing directional control valves with a single variable displacement pump, it is necessary to incorporate shuttle valves into the system. These valves compare the load sense signals from the working sections and output the highest load signal to the load sense pressure-flow compensator, which controls the pump to serve the highest loaded function.

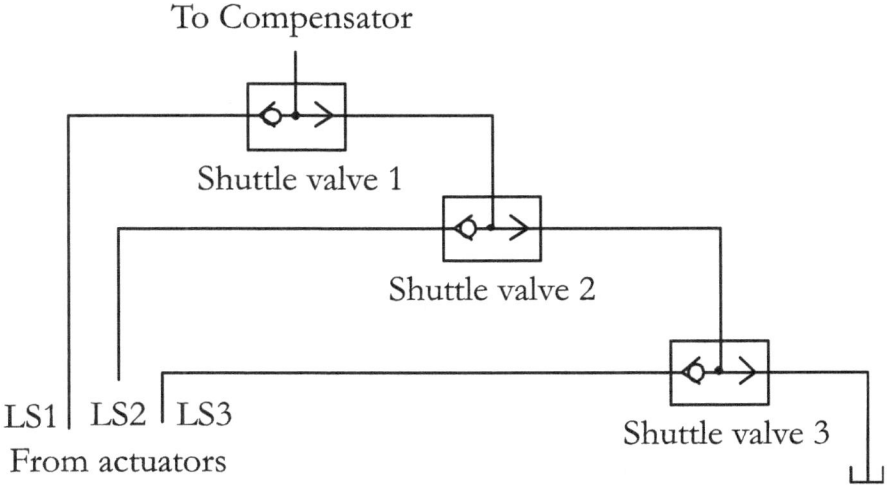

Figure 2.11 | Shuttle valves in a load sensing system with multiple functions

Typical Controllers for Variable Axial Piston Pumps

Many devices are available as controllers for the displacement and pressure control of variable-displacement axial piston pumps. Some of the typical pressure controllers are given below:

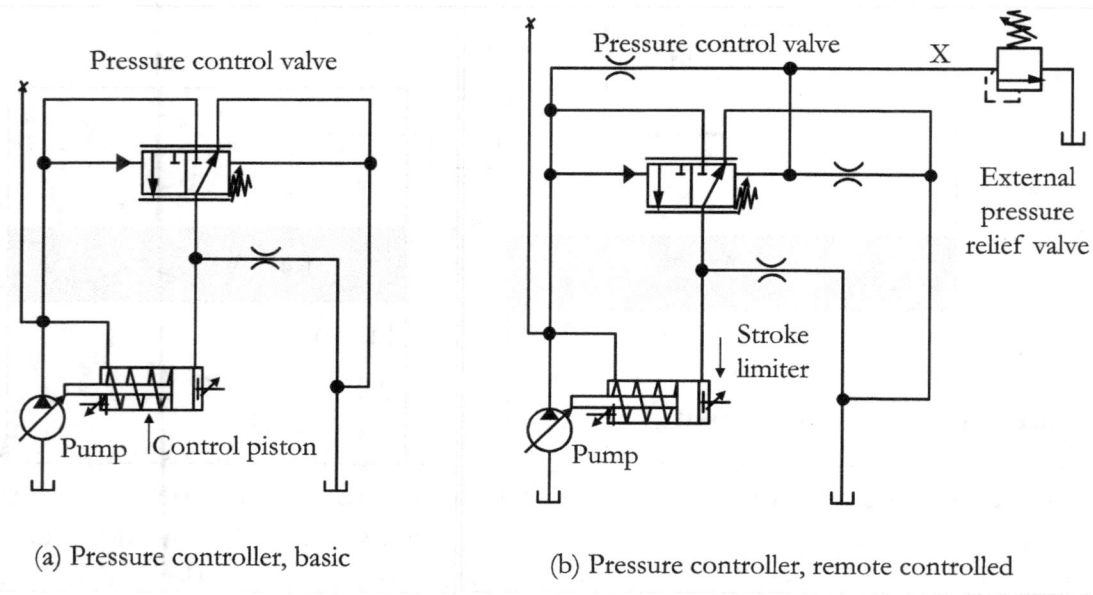

(a) Pressure controller, basic

(b) Pressure controller, remote controlled

Figure 2.12 | Typical pressure controllers

Pressure Controllers, Basic

A pressure controller limits the maximum pressure at the pump outlet within the control range of the variable pump. The variable pump only supplies as much hydraulic fluid as the consumers require. If the working pressure exceeds the pressure command value at the pressure valve, the pump will adjust to a smaller displacement to reduce the control differential. A simplified schematic of the pressure controller is given in Figure 2.12(a).

Pressure Controllers, Remote-Controlled

The remote-controlled pressure controller has a fixed setting Δp value. A separately connected pressure relief valve at port X enables the pressure controller to be remote-controlled. A simplified schematic of the pressure controller is given in Figure 2.12(b).

Flow Controller

Figure 2.13 displays a flow controller.

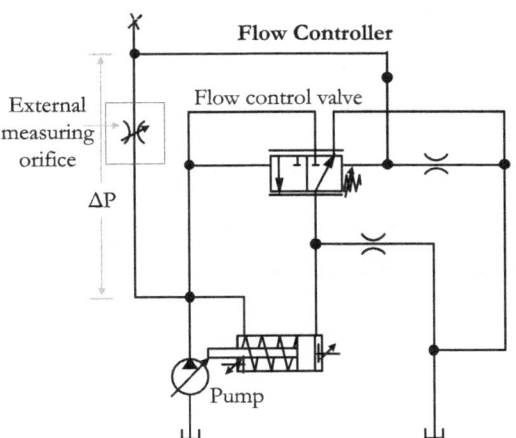

The purpose of the flow controller is to regulate the pump's output to match the volume needed by the connected actuator (not shown). This is achieved by measuring the flow through an external orifice located between the pump and the actuator and adjusting the pump's flow rate accordingly. The flow controller monitors the pressure before and after the orifice and maintains a consistent pressure differential (Δp) in order to control the flow. When there is an increase in pressure differential, the swash plate angle of the pump also increases, resulting in more flow. Conversely, the swash plate angle decreases if the pressure differential decreases, leading to less flow until the valve is balanced. The spring chamber at the right-hand side of the pressure-flow compensator spool can drain fluid through a small orifice into the tank.

Figure 2.13 | Flow controller

Flow Controller with a Remote-controlled Pressure Controller

A flow controller with a remote-controlled pressure controller is in Figure 2.14.

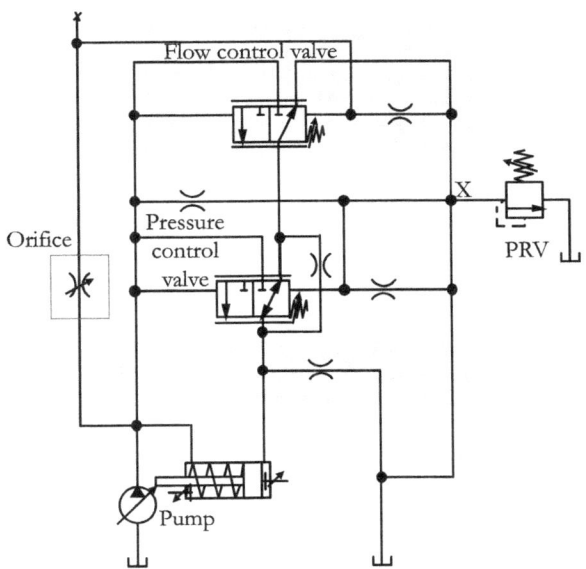

Flow Controller with Remote-controlled Pressure Controller

The pressure-flow controller combines a flow controller and a pressure controller. The flow controller regulates the pump's output to match the required volume of the connected actuator. This is done by measuring the flow through an external orifice between the pump and actuator and adjusting the flow rate accordingly. The pressure controller limits the maximum pressure at the pump outlet. If the working pressure exceeds the pressure command value at the pressure valve, the pump will regulate to a smaller displacement to reduce the control differential. A separate pressure-relief valve controls the pressure control, which can override the flow control.

Figure 2.14 | Pressure controller, remote controlled

Chapter 3 | Working of Load Sensing Systems

A typical load sensing system for controlling a hydraulic actuator consists of a variable-displacement pump, a compensator block, and a load-sensing closed-center DC valve. The system slips into various modes of operation during its cycle of operation. The following sections explain the operation of the load sensing system in its: (1) initial position, (2) low-pressure standby mode, (3) load-sensing mode, and (4) high-pressure standby mode.

Initial Position

Figure 3.1 shows the initial position of the load sensing system. The 290-psi and 4350-psi springs of the compensator force the pressure-flow compensator spool and high-pressure compensator towards their respective left-hand envelopes when there is no pressure in the system. The normal positions of the two spools provide for the fluid a direct passage from the swash plate control piston to the tank. As there is no fluid pressure acting on the control piston of the pump, the swashplate of the pump is forced to move to its maximum angle position. The pump is ready to produce the maximum flow in this initial position.

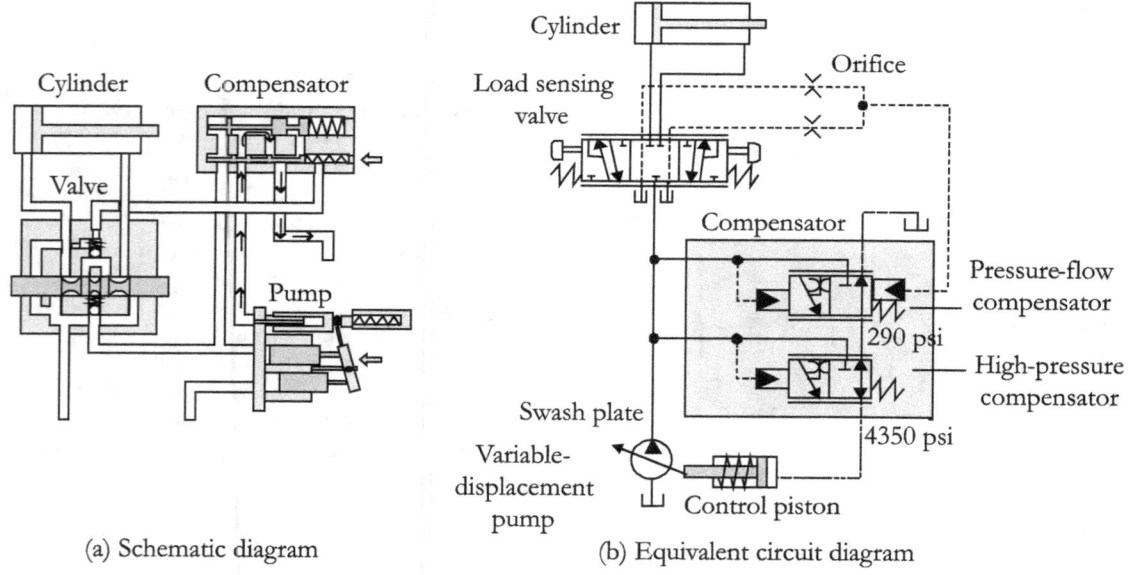

(a) Schematic diagram   (b) Equivalent circuit diagram

Figure 3.1 | A schematic diagram and a corresponding circuit diagram of a load sensing system in its initial position

Low-Pressure Standby Mode

The low-pressure standby mode is shown in Figure 3.2. Assume that the flow to the hydraulic actuator is blocked by the closed-center directional control valve when the pump is switched on. However, the pump flow enters the compensator block and acts on the left-hand side of the pressure-flow and high-pressure compensator spools.

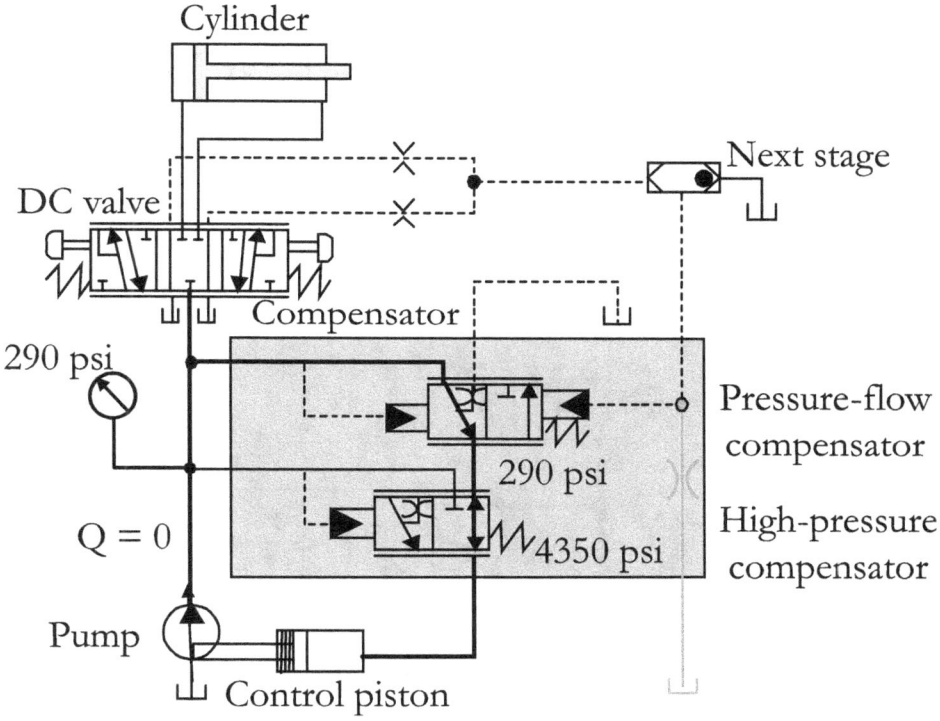

Figure 3.2 | A load sensing system in the low-pressure standby mode

When the pressure reaches 290 psi, the pressure-flow compensator spool moves to the right against the low-pressure spring, and hence, the flow is directed to the swash plate control piston. This flow causes the swash plate to de-stroke and makes the pump deliver a minimum flow at low pressure to the idling system. This position of the system is the low-pressure standby mode. In this mode of operation, the pump provides only enough flow to make up for the internal leakage in the system at low pressure.

It is important to mention that the fluid from the spring chamber of the pressure-flow compensator can be released into the tank through a small orifice in the compensator or directional control valve. However, there will not be a significant pressure drop when the signal is sent to the load sensing port of the compensator because the orifice is very small.

Load Sensing Mode

Assume that the spool of the load sensing directional control valve is shifted to the left-hand-side envelope, as shown in Figure 3.3. The flow is now directed to the cylinder and the right-hand side of the pressure-flow compensator spool. The pilot line senses the pressure as demanded by the load. The pilot pressure and the force of the 290-psi spring move the pressure-flow compensator spool to the left and drop the pressure from the swash plate control piston.

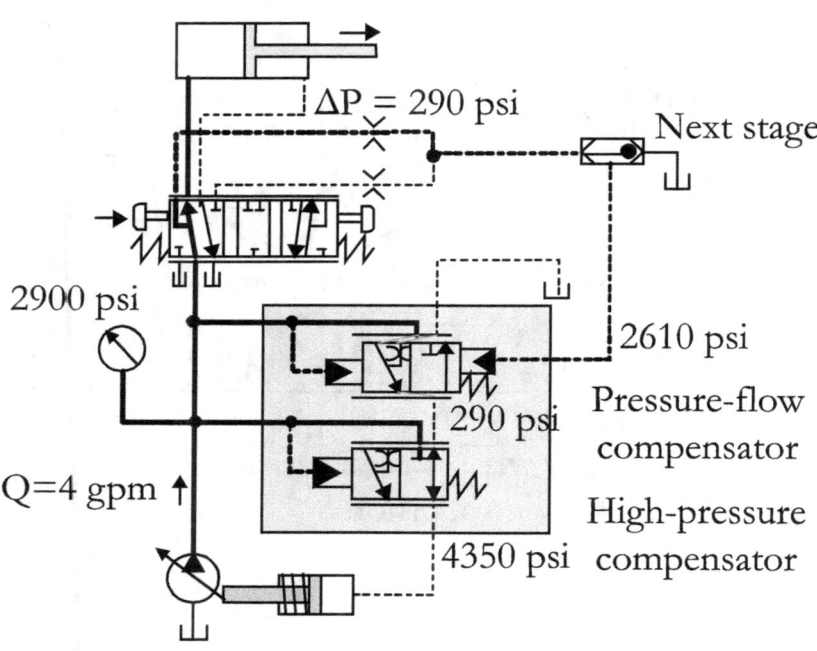

Figure 3.3 | A load sensing system when the DC valve is just actuated and sensing the load

The swash plate moves to a position with a greater angle, and the pump begins to produce more flow. The pump now delivers the required flow at a slightly higher pressure to meet the load. The proportional movement of the pressure-flow compensator spool controls the pump flow as per the load requirement. It can be noted that there is always a pressure difference of precisely 290 psi acting on the pressure-flow compensator spool. The pump only provides the fluid at a pressure of 290 psi above the actual working pressure. In the load sensing mode, the pump adjusts to the system's varying pressure and flow demands.

High-pressure Standby Mode

Eventually, the cylinder reaches the end of its stroke, and the system pressure goes up. The pressures on both ends of the pressure-flow compensator spool remain equal, as shown in Figure 3.4. The 290-psi spring and the load pressure force the pressure-flow compensator spool to the right-hand side envelope position.

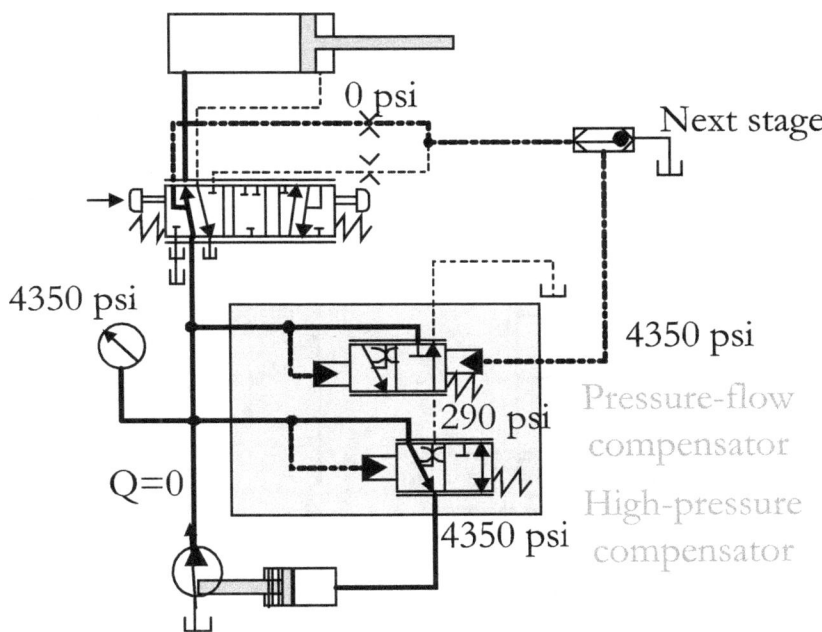

Figure 3.4 | A load sensing system in the high-pressure standby mode

When the load pressure reaches the required maximum level (say 4350 psi), the high-pressure compensator spool moves proportionately to the left-hand-side envelope position and directs fluid to the swash plate control piston. The piston moves the swash plate to its near-zero angle position, and the pump stops producing flow. This position is called the high-pressure standby mode. It provides only enough flow to make up for internal leakage in the system at high pressure.

A Load Sensing System with Multiple Actuators

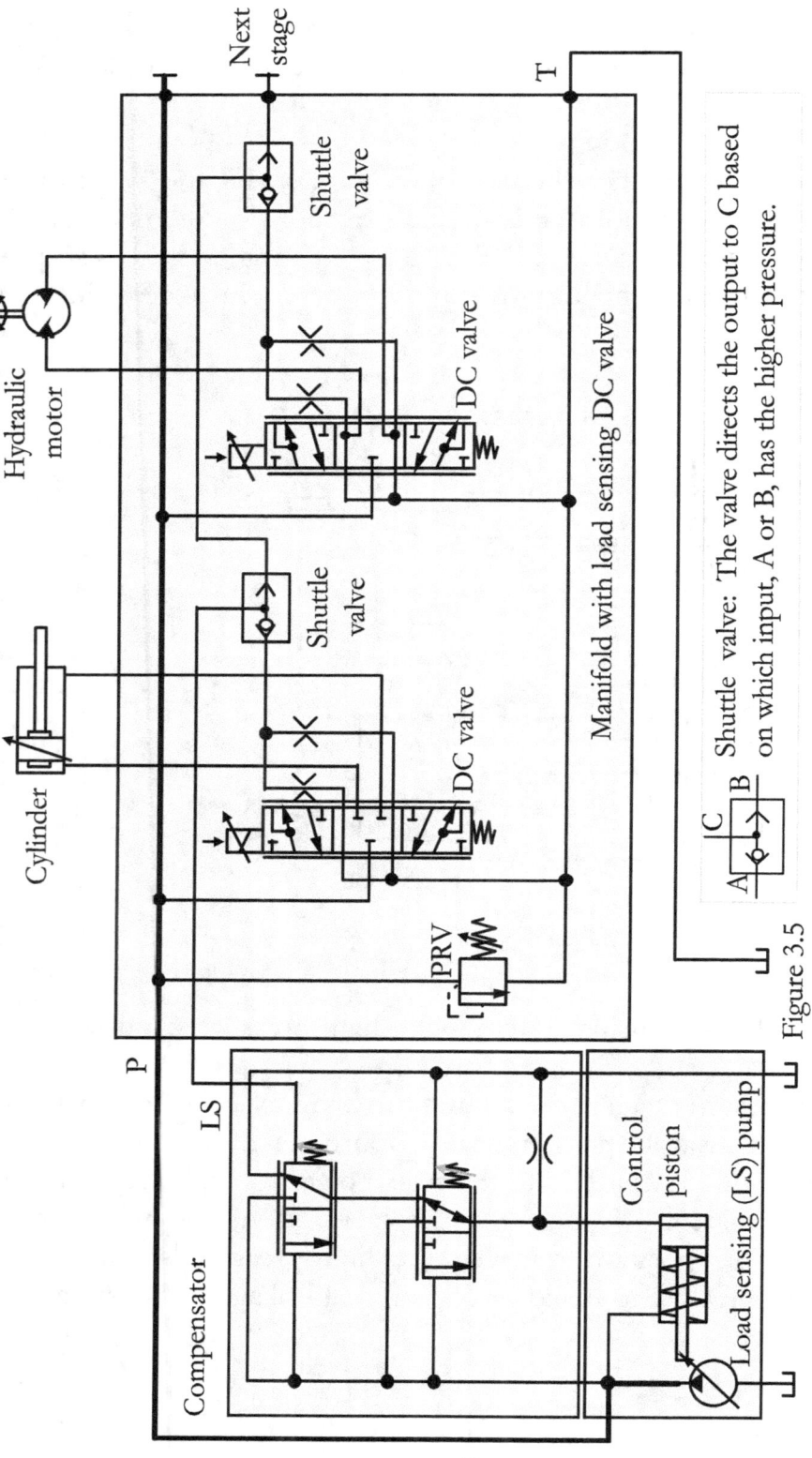

Figure 3.5

Explanation : Figure 3.5 shows a hydraulic system with multiple actuators that uses load sensing technology. This system comprises a load sensing hydraulic pump, a double-acting cylinder, and a hydraulic motor. The load-sensing pump has a compensator with a load sensing (LS) port, which receives the higher pressure of the load through shuttle valves. This pump only supplies the flow and pressure that the system needs, making it more efficient. If you need to add more actuators to the system, you can do so with additional shuttle valves.

Example 3.1 | Calculate the power required to drive a constant displacement pump in a conventional hydraulic system for producing a flow of 20 gpm at a pressure of 2000 psi.

If the flow required for moving a load is only 5 gpm at 2000 psi, calculate the power wasted in the system.

What is the power saving if the conventional system is replaced by a load sensing system with a variable-displacement pump and compensator with a pressure setting of 200 psi?

Neglect mechanical losses due to friction.

Solution

System flow $=20$ gpm
Pressure $=2000$ psi

Power $=20 \times 2000 / 1714 .23.3$ hp

As only a 5-gpm flow is required to move the load, the remaining 15-gpm flow is returned to the reservoir.

Therefore, power required to move the load
$$=5 \times 2000 / 1714 \text{ hp} = 5.83 \text{ hp}$$

Power wasted in a conventional system
$$=23.3 - 5.83 = 17.47 \text{ hp}$$

Power required moving the load in the load sensing system
$$=5 \times (2000 + 200) / 1714 = 6.4 \text{ hp}$$

Power saving using the load sensing system
$$=23.3 - 6.4 = 16.9 \text{ hp}$$

Chapter 4 | Electro-hydraulic Load Sensing Systems

Designers of hydraulic systems aim to enhance efficiency by introducing new technologies. The first-generation hydro-mechanical load sensing systems, equipped with pressure limiters, were created to improve the controllability and efficiency of hydraulic systems. Further advancements in these areas have been made possible by integrating electronic components into standard hydro-mechanical load sensing systems, forming electro-hydraulic load sensing systems.

For example, an electronic pressure transducer can be installed on the pump outlet line to instantaneously measure the pressure at the pump outlet. This pressure can then act on one side of the load sensing regulator. Further, a proportional electronic regulator can measure the load pressure, acting on the other side of the load sensing regulator. This results in the pump outlet pressure and flow being proportional to the differential signals that act on the load sensing regulator. Therefore, the load sensing function can be achieved electronically by obtaining the instantaneous measurement of two pressure transducers.

In addition, when an angular transducer is integrated to measure the pump swashplate's position, it becomes possible to accurately determine the pump's torque. As a result, the hydraulic system can be optimized for torque control, which enables the engine power to be utilized more efficiently.

Figure 4.1 displays a typical electro-hydraulic load sensing valve block diagram.

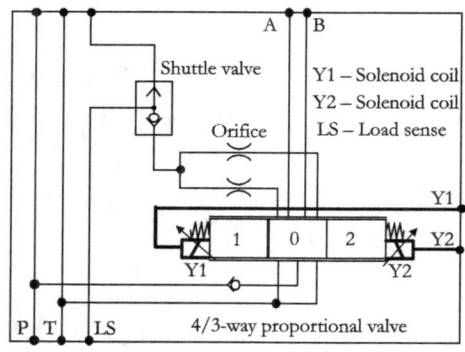

Figure 4.1 | A typical electro-hydraulic load sensing valve system

Chapter 5 | Advantages and Disadvantages of Load Sensing Systems

Load sensing hydraulic systems are designed to conserve energy by adjusting the pressure to meet the power requirements. These systems offer a significant benefit as they ensure great efficiency overall.

By regulating both the pressure and flow in a hydraulic system, there can be a noticeable decrease in losses, an increase in circuit efficiencies, and a longer lifespan for the entire system.

A load sensing system with modular valves has a simpler design with less piping, fewer components, and less control wiring, which makes the system more straightforward. Additionally, onboard electronics and bus-based control architecture make commissioning, maintenance, and troubleshooting much more manageable. Moreover, these valves can incorporate advanced capabilities that promote safer and easier operation.

Utilizing a load sensing hydraulic system presents several advantages. Below are some critical benefits of using this system:
- generates lower heat levels
- use less power
- consumes less fuel
- cooler system operation
- longer component life
- greater reliability
- better load control
- makes better use of available horsepower
- operate multiple cylinders with only one pump

Disadvantages: The cost of a load sensing system may vary depending on the components used, which could be a potential drawback. Load sensing systems with valve compensators that have an oscillatory nature can have a negative characteristic of triggering unwanted resonance effects. These effects can be amplified by dynamically varying loading conditions. Another disadvantage is that the actuator may operate slowly when the pump is signaled to increase flow.

Chapter 6 | Applications of Load Sensing Systems

The load sensing feature can be an ideal hydraulic system requirement when the system has one or more of the following characteristics: (1) system pressure peaks frequently, (2) higher level of energy consumption, (3) the possibility of excessive heat development in the system, and (4) one pump need to operate multiple circuits with variable pressure and flow conditions.

Load sensing valves bring flexibility to the control of hydraulic systems and enhance reliability as there are fewer components and reduced plumbing.

The load sensing system provides consistent operating speed for the machine, regardless of the load. It also has a pressure compensator for flow-sharing performance, ensuring controllability during multiple-function operations.

The load sensing systems are versatile and can be used in many hydraulic systems. Notably, the principle of load sensing is widely used in mobile hydraulic systems, including forestry, agriculture, and construction. These systems are suitable for both medium-pressure and high-pressure applications.

Load sensing controls enable accurate positioning of cylinders and precise speed control of motors. This makes it possible to perform operations in a crane, such as equipment leveling and stabilization using outriggers and boom extension, retraction, and rotation with exceptional accuracy.

The load sensing systems are also used in many other applications, including tractor systems, lifts, trucks, marine generator drives, machine tool systems, presses, and vehicle suspension systems. Load sensing has been used on many of the following machine applications:

- Tractor systems	- Front-end loaders	- Propulsion ground drives
- Backhoes	- Log splitters	- Salt spreader trucks
- Cable winches	- Machine tool systems	- Truck-mounted augers
- Concrete trucks	- Presses	- Vehicle suspension systems
- Cranes	- Scissor lifts	- Garbage compaction trucks
- Dump bed trucks	- Drill rigs	- Marine generator drives

7 | Test Your Knowledge

1. A(n) _____ system provides only the pressure and the flow as required by a hydraulic system.

2. A pump compensator of a load sensing system consists of a _____ compensator spool that works against a low-pressure spring and a high-pressure compensator spool that works against a high-pressure spring.

3. In the _____ mode of a load sensing hydraulic system, its pump produces only enough flow to the idling system to make up for the internal leakage in the system.

4. In the _____ mode of a load sensing system, its pump automatically adjusts itself to the varying pressure and flow demands of the system.

5. In the _____ mode of a load sensing hydraulic system, its pump produces only enough flow against the high-pressure load to make up for the internal leakage in the system.

[Choose from: cartridge valve, high-pressure standby, load sensing, low-pressure standby, pressure-flow, servo valve]

8 | Objective Type Questions
1. A load sensing control with a variable-displacement pump can:
a) compensate the pump for leakage flows
b) set the maximum pressure in the associated system
c) assist the system in counterbalancing the system load
d) match pump output flow and pressure to system demand

2. Mark the underline incorrect statement.
a) A load sensing system can be configured using a load sensing type pump, compensator, and a load sensing type directional control valve.
b) The load sensing compensator combines a pressure-flow compensator with a high-pressure compensator.
c) A load sensing system promotes heat development in a system.
d) A load sensing system improves performance and energy efficiency.

9 | Review Questions

1. Draw the pressure-flow characteristic of a hydraulic system with a fixed-displacement pump delivering only half the flow to the load, showing the power utilized and power wasted.

2. Draw the pressure-flow characteristic of a hydraulic system with a variable-displacement pump delivering only half the flow at less pressure to the load, showing the power utilized and power wasted, if any.

3. Draw the pressure-flow characteristic of a load sensing hydraulic system with a variable-displacement pump delivering only half the flow at less pressure to the load, showing the power utilized and power wasted, if any.

4. Differentiate between the conventional and the load sensing hydraulic systems.

5. Explain the operation of a variable-displacement pump with a control piston as used in a load sensing system with a simple schematic.

6. Explain the working of the pump compensator of a load sensing system with a simple schematic diagram.

7. Can you explain the use of mechanical stop screws in a variable displacement pump's bias cylinder and displacement control cylinder?

8. Explain the working of a pressure controller for variable-displacement pumps.

9. Can you explain the importance of delta-P in a hydraulic system for elevating cylinders?

10. Can you elaborate on combining load sensing signals in a hydraulic system that serves multiple functions?

11. What is a load sensing hydraulic system?

12. Describe the working of a load sensing system with a simple circuit diagram.

13. Explain how the low-pressure standby mode is achieved in a load sensing hydraulic system.

14. Explain how a high-pressure standby mode is achieved in a load sensing hydraulic system.

15. What is the difference between the low-pressure and the high-pressure standby in a load sensing hydraulic system?

16. How is the high-pressure limiting achieved in a load sensing hydraulic system?

17. What are the advantages of using the load sensing feature in hydraulic systems?

18. Write a short note on the hydraulic applications of load sensing control.

10 | References

1. Article on: 'Agricultural load-sensing Hydraulic systems', H. Richard Jarboe, published by the American Society of Agricultural and Biological Engineers, St. Joseph, Michigan. www.asabe.org

2. Article on: 'Electro-hydraulic load sensing with alternating control', Dipl.-Ing. Björn Grösbrink, Institut fur Landmaschinen und Fluidtechnik, Technische Universitat, Braunschweig

3. Article on: 'Hydraulic Closed-Center Circuit with Load Sensing', Hydraulic Valve Troubleshooting & Repair, www.valvehydraulic.com

4. Article on: 'Understanding hydraulic load sensing control', Insider Secrets to Hydraulics, http://www.industrialhydrauliccontrol.com/

5. Catalog 'VP120 Load-Sense Directional Control Valve, Motion Hydraulic Valves', HY14-2008/US, Parker Hannifin Corporation, Hydraulic Valve Division, 520 Ternes Avenue Elyria, Ohio 44035 USA

6. Documents on 'Axial Piston Variable Pump A4VSO Series 1, 2 and 3, Operating Instructions', RE 92050-01-B/03.08, and 'Axial piston variable pump A15VSO, A15VLO Series 11' Bosch Rexroth AG Hydraulics, Axial Piston Units, An den Kelterwiesen, Horb, Germany

7. Document on 'Pressure and flow control system Type SYDFE1 series 2X, 3X with external control electronics VT 5041-3X, Operating instructions', Bosch Rexroth AG, Industrial Hydraulics, Zum Eisengiesser 1, 97816 Lohr a. Main, Germany

8. Document on: 'Load sensing hydraulics', Barko Hydraulics. LLC

9. Document on: 'Load Sensing Systems - Principle of Operation', Eaton Corporation, Hydraulics Division, Eden Prairie, MN, USA

10. Document on: 'Load-sensing hydraulic system', Aebi Schmidt Holding AG, Thurgauerstrasse, Zürich

11. Document on: 'Mobile Hydraulics: Modular Load-Sensing Valves', Bosch Rexroth Corporation, Mobile Hydraulics, Wooster, OH, USA

12. Manual 'A10VO Variable Displacement Piston Pump, Technical Information Manual, Module 3A', Bosch Rexroth Canada, 16.06.2006 ı Revision 2.0

13. Technical Information Directional Control Valve ECO 80, BC199786485316en-000301, Danfoss Power Solutions (US) Company, 2800 East 13th Street Ames, IA 50010, USA

Fluid Power Educational Series Books

1. Pneumatic Systems and Circuits -Basic Level (In the SI Units)
2. Industrial Pneumatics -Basic Level (In the English Units)
3. Pneumatic Systems and Circuits -Advanced Level
4. Electro-Pneumatics and Automation
5. Design of Pneumatic Systems (In the SI Units)
6. Design Concepts in Pneumatic Systems (In the English Units)
7. Maintenance, Troubleshooting, and Safety in Pneumatic Systems
8. Industrial Hydraulic Systems and Circuits -Basic Level (In the SI Units)
9. Industrial Hydraulics -Basic Level (In the English Units)
10. Hydraulic Fluids
11. Hydraulic Filters: Construction, Installation Locations, and Specifications
12. Hydraulic Power Packs (In the SI Units)
13. Power Packs in Hydraulic Systems (In the English Units)
14. Hydraulic Cylinders (In the SI Units)
15. Hydraulic Linear Actuators (In the English Units)
16. Hydraulic Motors (In the SI Units)
17. Hydraulic Rotary Actuators (In the English Units)
18. Hydraulic Accumulators and Circuits (In the SI Units)
19. Accumulators in Hydraulic Systems (In the English Units)
20. Hydraulic Pipes, Tubes, and Hoses (In the SI Units)
21. Pipes, Tubes, and Hoses in Hydraulic Systems (In the English Units)
22. Design of Industrial Hydraulic Systems (In the SI Units)
23. Design Concepts in Industrial Hydraulic Systems (In the English Units)
24. Maintenance, Troubleshooting, and Safety in Hydraulic Systems
25. Hydrostatic Transmissions (HSTs) (In the SI Units)
26. Concepts of Hydrostatic Transmissions (In the English Units)
27. Load Sensing Hydraulic Systems (In the SI Units)
28. Concepts of Load Sensing Hydraulic Systems (In the English Units)
29. Electro-hydraulic Proportional Valves
30. Electro-hydraulic Servo Valves
31. Cartridge Valves
32. Electro-hydraulic Systems and Relay Circuits
33. Practical Book: Pneumatics - Basic Level
34. Practical Book: Electro-pneumatics - Basic Level
35. Practical Book: Industrial Hydraulics – Basic Level
36. Programmable Logic Controllers and Programming Concepts
37. Compressed Air Dryers
38. Hydraulic Circuits – Identification of Components and Analysis

For more details, please visit: **https://jojibooks.com**

About the Author

Joji Parambath is an accomplished expert in Pneumatics, Hydraulics, and PLC with an extensive 25-year background in the field. Over the course of his career, he has trained a multitude of professionals from diverse industries, as well as faculty members and engineering students.

Joji is the primary faculty member at Fluidsys Training Centre in Bangalore, India, offering comprehensive training in Pneumatics and Hydraulics. He has authored an impressive 39 books on the subject matter, all designed to convey knowledge on Pneumatics and Hydraulics in a simplistic and easy-to-understand manner.

Joji attributes the creation of his book series to the active engagement and valuable suggestions of his trainees during the training programs. He would like to extend his gratitude towards them.

28th July 2023

www.ingramcontent.com/pod-product-compliance
Lightning Source LLC
Chambersburg PA
CBHW080109010626
45794CB00015B/3338